How Gorilla Closely related to Human?

(Wisdom of the Jungle)

Nsesheye Nkota

ISBN:
ISBN-B9781717862808

DEDICATION

To all those who devort their lives to ensure the security and existence of creature gorilla in our world now and for the benefit of future generations in order to glorify Sir God for his creation

CONTENTS

ABSTRACT

A gorilla is a powerfully built great ape, aggressive looking man with a large head and short neck, found in the forests of central Africa. They're peaceful, family oriented, plant-eating animals. Many people like to compare gorillas with humans. Scientists have shown that gorillas live in complex social groups, display individual personalities, make and use tools, and show emotions like grief and compassion. We shouldn't be surprised though - Gorillas are one of our closest living relatives, and they share at least 95% of their DNA with humans. British anthropologist Robin Dunbar criticized this basic philosophy, He argued this clearly in a 2008 article titled Why Humans Aren't Just Great Apes. Within this article Dunbar claims that humans are different from great apes in one critical respect: our imagination (Dunbar, 2008). Prof Jonathan Marks is in line with Dunbar in his post at website PopAnth called "Are we apes? No, we are humans". Like all great apes (except humans), gorillas require rain forests to make their living, and the forest depends upon them, too. The gorilla's fibrous scat acts as rich fertilizer for the forest, and seedlings sprout from it rapidly, making these animals important forest regenerators. In this book, you are going to see how Gorilla is closely related to human and now you will start to understand the Wisdom of the Jungle.

1.0 INTRODUCTION

Human Gorilla is highly related to people than thought as human is highly evolved primates. It should be noted that, Gorillas are fruit- eating anthropoid Apes (Rogers et al., 2004; Doran and McNeilage, 2001; Yamagiwa et al., 2003) and largest and strongest of all apes. Gorillas also eat leaves, stems, seeds, and roots. Despite being immense in size, as you can imagine, most of their diet is made up by foliage and fruit, with the small exception of some species that sometimes eat small ants or insects. Due to the low nutritional value of their main food, gorillas have to consume a lot of foliage and fruits, so they spend most of their days foraging for food.

Like humans, gorillas reproduce slowly, giving birth to only one baby at a time and then raising that infant for several years before giving birth again. This slow reproduction rate makes gorillas especially vulnerable to any population declines.

Habitat destruction is a problem across their central African range states namely Angola, Cameroon, Central African Republic, Congo, Democratic Republic of Congo (DRC), Equatorial Guinea, Gabon, Nigeria, Rwanda, Uganda. Gorillas are also still killed for the bushmeat trade. That trade has helped spread the Ebola virus, which is deadly to both gorillas and humans. Efforts to protect gorillas are often hampered by weak law enforcement and civil unrest in many places where gorillas live. A careful follow up need to be initiated to rescue these amazing creatures, don't forget we are all responsible to their safety, it is a gift from Sir God.

The following map addresses the gorilla priority area to be taken care of.

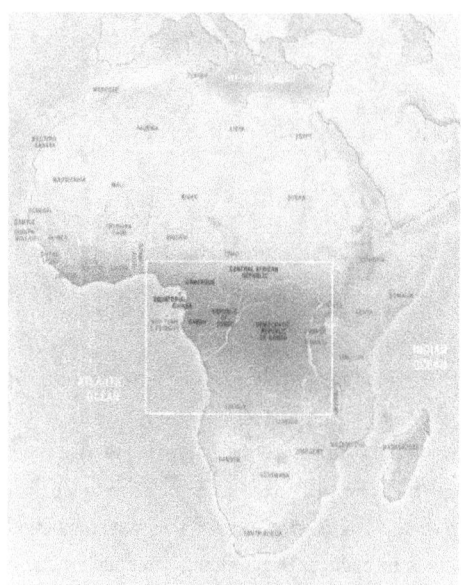

GORILLA PRIORITY PLACE
"Congo Basin"

Habitat

Major-habitat-type
Tropical and Subtropical Moist
Broadleaf Forests

Biogeographic realm
Afrotropical

Range States
Angola, Cameroon, Central African
Republic, Congo, Democratic Republic
of Congo (DRC), Equatorial Guinea,
Gabon, Nigeria, Rwanda, Uganda

Ecological Region
Guinean Moist Forests, Congolian Coastal Forests, Cameroon Highlands Forests,
Northeastern Congo Basin Moist Forests, Central Congo Basin Moist Forests, Western
Congo Basin Moist Forests, Albertine Rift Montane Forests.

In understanding how Gorilla Closely related to Human, the main areas are going to be
discussed namely social behaviours, Genetic differences, Physical characteristics of
human skul, skeleton and Digestive system related to diet and eating habit.

1.1 CHAPTER 1

1.2 What is this creature Gorilla?

A gorilla is a powerfully built great ape, aggressive – looking man with a large head and short neck, found in the forests of central Africa. It is the largest living primate. But the truth is, they're peaceful, family oriented, plant-eating animals that live in complex social groups. They are the largest of all primates—the group of animals that includes monkeys, lemurs, orangutans, chimpanzees, and humans. The life span of gorilla is normally 35 to 40 years although zoo gorillas live 50 to 60 years if properly taken care of.

Many people like to compare gorillas with humans, are they the same or is that a Cousin from another Nephew?. According to scientists, "our closest animal relatives are the great apes: chimpanzees, orangutans and gorillas. About 98% of the DNA in your genes is exactly the same as in chimpanzees, making you as closely related to a chimp as horses are to zebras. Chimps and humans share a common ancestor, who was probably swinging through the trees about 5 million years ago. Many other species of ape around at the same time eventually became extinct".[1]

---

[1]

http://whoami.sciencemuseum.org.uk/whoami/findoutmore/yourgenes/wheredidweco mefrom/whatareourclosestanimalrelatives

3

Author: Nsesheye Nkota

According to an evolutionary anthropologist Cadell, L(2013), "The Great (Ape) taxonomy debate", he argued that, I don't think it is this straightforward. Whether we are great apes (lumping) or aren't great apes (splitting) doesn't just affect research and theory, but it also affects how we conceptualize what it means to be human. So what is the scientific evidence and reason behind lumping us in with the great apes? The main justification for grouping humans and great apes together is anatomical and morphological. We share the same 'Y5' pattern (five cusps or raised bumps arranged in a Y-shape), a rotating shoulder, no tail, posteriorly positioned scapula, fused caudal vertebrae, and a large and complex brain (Marks, 2009).

To me, the most important of these anatomical and morphological similarities is the "large and complex brain." It is true that the great apes have larger brain to body size ratios than all other primates. And it is this large brain size that allows the great apes to accomplish incredible intellectual accomplishments both in the wild and in laboratory settings. However, the human brain is at least three times the size of any great ape brain and Most importantly, the human brain has enabled levels of communication and intelligence unparalleled in the history of life on Earth (Cadell, L(2013).

Biological anthropologist Jonathan Marks has recently explored the first of these reasons in great detail (e.g., Marks, 2009; Marks 2012). Marks claims that by lumping our species in with great apes, researchers create a human evolutionary framework that begins by assuming we never really became human. It also implicitly rejects an important Darwinian theoretical approach by focusing on descent, as opposed to divergence (Marks, 2009).British anthropologist Robin Dunbar agrees with this basic philosophy, and stated this clearly in a 2008 article titled Why Humans Aren't Just Great Apes. Within this article Dunbar claims that humans are different from great apes in one critical respect: our imagination (Dunbar, 2008). I would contend that there are several more critical differences between the human mind and the great ape mind, but I do agree with Dunbar's approach of stressing the ways in which humans are divergent from the great apes.

A further comparison based on Physical characteristics of their skuls versus that of human skul, skeleton and Digestive system related to diet and eating habit, social behaviours, and Genetic differences will give you a clear undestanding about so many raised questions in you mind.

1.3 Gorilla skull Vs Human skull.

Skull: the structural shape of gorilla and human are similar. They have equivalent bones of the same names though of different shapes and proportions.

Human Skull

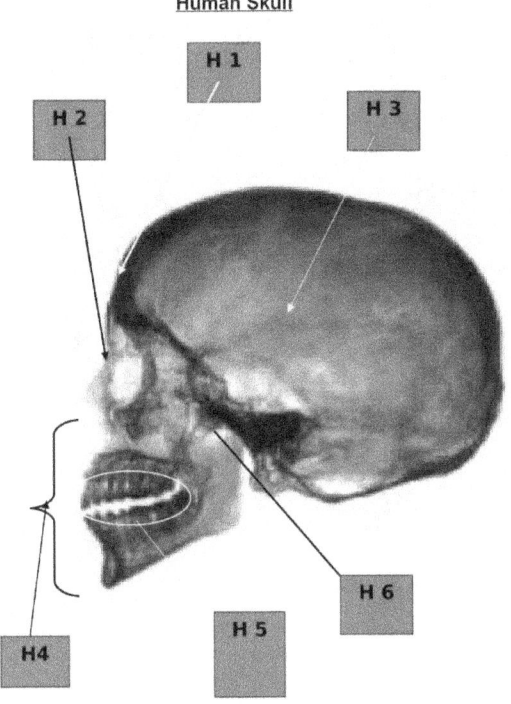

Gorilla Skull

G4 G 1 G 2 G3

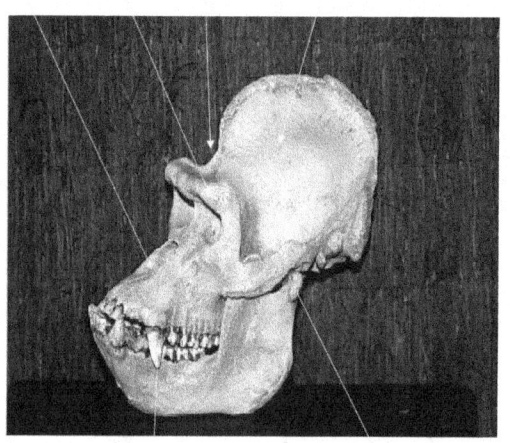

G5 G6

Human skull	Gorilla skull
1. High forehead.	1. No forehead
2. Small brow ridge.	2. Large and distinctive brow ridge
3. Large brain case than in gorilla	3. Small brain case (for smaller brain)
4. Flattered face rather than Projecting	4. Face projecting forward.
5. Teeth include small molars vericle incisors)	5. Large Canine teeth.
6. Smaller zygomatic arch than in gorilla	6. Larger zygomatic arch (muscles used for chewing action-ie Surface)

1.4 Gorilla skeleton Vs Human skeleton

Since their skulls is just one part of their skeletons, let us see their skeleton. Actually their skeleton have the same badiac structure. That is to say. Once you lable the bones of human skeleton, probably you can also lable the bones of gorilla skeleton though their shapes and proportions are different. See their diagram below and you will identify their differences.

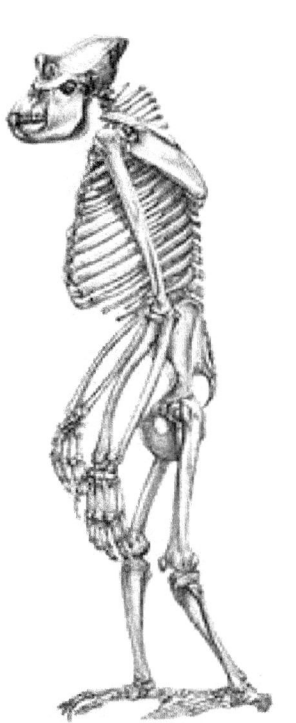 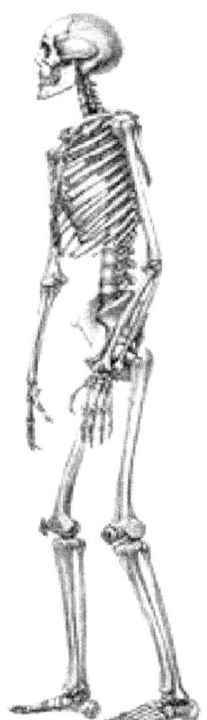

Gorilla Skeleton   Human Skeleton

Author: Nsesheye Nkota

Human skeleton	Gorilla skeleton
1.shapes and proportions of their skuls as displayed above)	1. check above.
2.Has S-shaped vertebral column.	2. Has bow-shaped
3. Human Legs are proportional longer than gorilla legs	3. Gorilla arms are proportional longer and strong than human for climbing trees,lifting, breaking, and squeezing health objects.
4. Human feet have 5 toes aligned with each other.	4. Gorilla feet have opposable large toe.
5. Human fingers are straight.	5. Gorilla fingers are curved
6. Human has proportionally large thumb than in gorilla.	6. Gorilla has proportionally Small thumb

It must be noted that, though gorillar can walk by using two legs, sometimes they also move on all four quadrupedal gaits.(i.e two hands and two legs.

1.5Social behaviors of Gorillas

Gorillas are highly social, relatively non-territorial and live in groups (called troops) that usually consist of 1-4 adult males, some juvenile males, and several young and adult females.   The oldest and strongest adult male (called the silverback) is usually the dominant one of the troop and has exclusive breeding rights with the females.   Gorillas display individual personalities, make and use tools, and show emotions like grief and

compassion. They also spend a good deal of their time on the ground rather than in the trees, and will make new nests on the ground each night. (see Robbins et al., 2001; The Dian Fossey Gorilla Fund International, www.gorillafund.org). Young gorillas stay with their mothers until they build their own nests when they are about 3 years of age. Gorillas activities include traveling in seaching for food (Remis 1997; Bermejo 2004; Doran et al., 2004; McFarland 2007), Feeding on various types of vegetation, social interaction forexample grooming, reproduction and parenting.

Male gorillas mature at approximately 11 to 13 years of age while females at 10 to 12 years. Male and female Gorillas tend to move away from the troop where they were born when they mature into adults. In case a troop is having single adult male and adult females and their offspring, generally disperse to join other troops.

Gorilla infants depend on their mothers for survival. gorilla infants remain in contact with their mothers for the first six months. Male gorillas play less active role in caring for their offspring than female but in otherway adult males play a big role forexample The adult males within the troop protect young gorillas from infanticidal males outside the troop. Once it occurs aggressive behaviour between female gorillas within the troop, may be addressed and usually concern social access to the males who sometimes interveine in settling such disputes. Nevertheless, adult male usually help to socialize the offspring immediately after they leave their mothers for enough length of time especially after about 18 months and into their juvenile period normally 3 to 6 years. Adult male gorillas do not usually form bond of friendship with each other in mixed gender troops where competition in high hierarchy position and for dominates arise. Surprisingly males in all male troops socialize with each other by playing ang grooming.

1.6 Genetic comparison revealed.

One difference, the researchers discovered is a gorilla gene that probably helps the animal's skin grow a tough layer of keratin- a protein that makes up finger nails on their knuckles which aids in animal's distinctive knuckle walk. Human seems to lack this genetic variant. Humans also underwent the deletion of some genes involved in the synthesis of fatty acids. Some genetic changes related to dietary metabolism were identified in this project. These may have played a role relevant to the evolution of ape species. Great ape diets range from keeping strictly vegetarian to eating almost anything. Another is sperm genes, in humans a specific gene enables sperm to compete with those of other males. In gorillas, those genes are in active. Smith, T (2012) in the 'Journal Nature', said some genes tied to dementia and heart failure in humans appeared roughly the same in both humans and gorillas, but are not harmful to gorillas."if we could understand more about why those variants are so harmful in humans but not in gorillas, that would have important useful medical implications" he said.

The researchers on this study predict that more advanced, long-range sequencing and mapping technologies, and even longer-read sequencing, will assist in increasing knowledge on the evolutionary journey taken by the great apes and our human ancestors.

2.0 CHAPTER 2

2.1 How gorilla genome gave scientists an insight into human evolution?

The interesting news, from researchers is that genome sequence for the gorilla has already been completed. The scientists note that many of the genetic differences between humans and other apes were not recognized when their genomes were first compared. Areas of rapid structural change were still nebulous in those early draft genome assemblies. This made them difficult to compare and limited the discovery of the functional differences that distinguish humans from other apes.

The newest investigation provides the most comprehensive catalog of genetic variants that were gained or lost on different ape lineages. As this was the last genus of the living great apes to have its genome decoded.

According to Zev N. Kronenberg, Ian T et al (2018) "High-resolution comparative analysis of great ape genomes", revealed that they were able to compare the genome for chimpanzees, gorillas,humans and orangutans for the first time.The team analysed over 11,000 genes in chimpanzees, humans and gorillas for genetic changes in evolution. The results confirmed that the chimpanzee is our closest relative, but revealed that our genome is more closely resembles the gorilla genome--up to 15 per cent of our genome is closer to the gorilla.

Scally, A et al (2018) also suggested that, "The gorilla genome is important because it gives light on the time when our ancestors diverged from our closest evolutionary cousins," "It also lets us explore the similarities and differences between our genes and those of gorilla, the largest living primate. The findings not only revealed differences in the species as a result of millions years of evolutionary divergence, but also parallel changes over time. For example, genes related to hearing, sensory perception and brain developed showed accelerated evolution in all three species, particularly in humans and gorillas.

According to Christopher M. Hill, et al (2018) Scientists had suggested that the rapid evolution of human hearing genes was linked to the evolution of language, "Our results cast doubt on this, as hearing genes have evolved in gorillas at a similar rate to those in humans." Additionally, the researchers studied brain organoids - laboratory-grown tissues coaxed from stem cells of apes or humans and forming a simplified version of organ parts. These brain proxies were examined to try to understand how differences in gene expression during brain development in humans and chimps might account for chimps' smaller brain volume, which is three times less than human brain volume. There are also significant dissimilarities in cortical structures in human and chimp brains.The researchers observed in the organoids that certain genes, particularly those in cells that are like the progenitors of radial glial neurons, are down-regulated in humans compared to chimps. Those genes are more likely to have lost segments of DNA specifically in the human branch important in regulating their expression. The genomes also indicated that gorillas diverged from humans and chimpanzees around ten million years ago.

The scientists caution that the ape genomes and their work on them are not yet complete because the genome assemblies are still missing other larger, more complex structural variations that cannot yet be assembled.

BIBLIOGRAPHY

1.   Alberts, S.C. Atlmann, J., Brockman, D.K., Cords, M., Fedigan, L.M., Pusey, A.E., Stoinski, T.S., Morris, B.F., Strier, K.B., and Bronikowski, A.M.. (2013). Reproductive cessation patterns in primates reveal that humans are distinct. *Proceedings of the National Academy of Sciences.*

2.   Andrews, P. Harrison, T (2005) The last common ancestor of apes and humans. In D.E. Lieberman, R.J. Smith & J. Kelley, Editors, *Interpreting the Past: Essay' on Human, Primate, and Mammal Evolution in honor of David Pilbeam,* 103-121. Boston, Brill Academic Publishers, Inc.

3.   Andrews, P. (1995) Ecological apes and ancestors. *Nature 376, 555 - 556.*

4.   Andrews, P. and J.E. Cronin. (1982), "The Relationship of Sivapithecus and Ramapithecus and theEvolution of the Orangutan." Nature, 297:541-546

5.   Andrews, P., Groves, C.P & Horne, J.F.M (1975). The ecology of the Lower Tana River Flood Plain. *Journal of the The Last African Natural History Society I51, 1-31.*

6.   Andrews, I. (1973). The vegetation of Rusinga island. Journal of the Last African Natural History Society 142, 1—8.

16

7. Barks, S.K., Calhoun, M.E., Hopkins, W.D., Cranfield, M.R., Mudakikwa, A., Stoinski, T.S., Patterson, F.G., Erwin, J.M., Hecht, E.E., Hof, P., & Sherwood, C.C. (2015). Brain organization of gorillas reflects species differences in ecology. *American Journal of Physical Anthropology, 156*, 252-262.

8. Bermejo, M. 2004. Home-Range Use and Intergroup Encounters in Western Gorillas (Gorilla g. gorilla) at Lossi Forest, North Congo. American Journal of Primatology 64:223-232.

9. Bronikowski, A.M., Atlmann, J., Brockman, D.K., Cords, M., Fedigan, L.M., Pusey,A.E., Stoinski, T.S., Morris, B.F., Strier, K.B., and Alberts, S.C. (2011). Aging in the natural world: Comparative data reveal similar mortality patterns across primates. Science, 331, (1325) DOI: 10.1126/science.1201571.

10. Bronikowski, A.M., Cords, M., Alberts, S. Atlmann, J., Brockman, D.K., Cords, M., Fedigan, L.M., Pusey, A.E., Stoinski, T.S., Morris, B.F., Strier, K.B & Morris, W. (2016). Female and male life tables for seven wild primate species. Scientific Data 3.

11. Broom, R. S: Schepers, G.W.H (1946). The South African fossil ape-men. The Australopithecinae. *Transvaal Museum Memoir,* Pretoria 2.

12. Caccone, Adalgisa. (1989), "DNA Divergence Among Hominoids." Evolution, 43(5): 925-941

13. Cadell, L(2013), "The Great (Ape) taxonomy debate" retrieved [2018 July 22] through https://blogs.scientificamerican.com/guest-blog/the-great-ape-taxonomy-debate/

14. Caillaud, D., Ndagijimana, F., Giarusso, T., Vecellio, V. & Stoinski, T.S. (2104). Mountain gorilla ranging patterns: Influence of group size and group dynamics. American Journal of Primatology, DOI: 10.1002/ajp.22265

15. Colchero, F., Rau, R., Jones, O.R., Barthold, J.A., Conde, D.A., Lenart, A., Nemeth, L., Scheuerlein, A., Schoeley, J., Torres, C., Zarulli, V., Altmann, J., Brockman, D.K., Bronikowski, A.M., Fedigan, L.M., Pusey, A.E., Stoinski, T.S., Strier, K.B., Baudisch, A., Alberts, S.C., Vaupel, J.W.. (2016) The emergence of longevous populations. Proceedings of the National Academy of Sciences. 113(48): 7681-7690

16. Clark Howell, F. & Bourliere, F. (1963). African Ecology and Human Evolution. Chicago, Aldine

17. Clark Howell, F. (1965). Early Man. New York, Time/Life Books.

18. Chiarelli, Brunetto. (1985), "Chromosome and the Origin of Man." Hominid Evolution: Past,Present, and Future. Alan R. Liss Inc., New York: 397-400

19. Dave, M (2012 March, 08) Gorillas More Related to People Than Thought, Genome Says. Retrieved July 08, 2018 from https://news.nationalgeographic.com/news/2012/03/120306-gorilla-genome-apes-humans-evolution-science/

20. Diamond, Jared (1992), The Third Chimpanzee. HarperCollins Publishers Inc., New York

21. Doran, DM, McNeilage, A. 2001. Subspecific variation in gorilla behaviour. In: Robbins, M.M., Sicotte, P., Stewart, K.J., eds. Mountain Gorillas, Three Decades of Research at Karisoke. Cambridge University Press, Cambridge, UK. pp. 123-149.

22. Dunbar, R. (2008), **Why Humans Aren't Just Great Apes.** Ethnology and Anthropology, 3: 15-33.

23. Eckardt, W., Steklis, D., Steklis, N., Fletcher, A., Stoinski, T.S. & Weiss, A. (2015). Personality structure of wild Virunga mountain gorillas. *Journal of Comparative Psychology,* 129, 26-41.

24. Eckardt, W., Steklis, D., Steklis, N., Fletcher, A., Stoinski, T.S. & Weiss, A. (2015). Personality structure of wild Virunga mountain gorillas. Journal of Comparative Psychology, 129, 26-41.

25. Eckardt, W., Stoinski, T.S., Rosenbaum, S., Umuhoza, M.R., Santymire, R. (2016). Characterizing stress physiology in the Virunga mountain gorilla. Journal Conservation Physiology.

26. Eckardt, W., Fawcett, K. & Fletcher, A (2016). Weaned age variation in the Virunga mountain gorillas (Gorilla beringei beringei): Influential factors. Behavioral Ecology and Sociobiology, 70, 493-507.

27. Erikson, E H. (1973), Godkin Lectures at Harvard University.

28. Fawcett, K., G. Bush, A. Seimon, G. Picton Phillips, D. Tuyisingize, and P. Uwingeli. (2011). Long term changes in the Virunga Volcanoes. In The Ecological Impact of Long-Term Changes in Africa's Rift Valley. Ed. A.J. Plumptre. Nova Science Publishers, UK.

29. Galbany J., Stoinski TS, Abavandimwe D, Breuer T, Rutkowski W, Batista NV, Ndagijimana F & McFarlin SC (2016). Validation of two independent photogrammetric techniques for determining body measurements of gorillas.American Journal of Primatology. DOI: 10.1002/ajp.22511.

30. Galbany J., Imanizabayo O, Romero A, Vecellio V, Glowacka H, Cranfield MR, Bromage TG, Mudakikwa A, Stoinski TS & McFarlin SC (2016) Tooth wear and feeding ecology in mountain gorillas from Volcanoes National Park, Rwanda. American Journal of Physical Anthropology. DOI: 10.1002/ajpa.22897.

31. Galbany, J., Abavandimwe, D., Vakiener, M., Eckardt, W., Mudakikwa, A., Ndagjimana, F., Stoinski, T.S., McFarlin, S.C. 2016. Body growth and life history in wild mountain gorillas (Gorilla beringei beringei) from Volcanoes National Park, Rwanda. American Journal of Physical Anthropology. 1-20.

32. Goodman, Morris et al. (1990), "Primate Evolution at the DNA Level and a Classification ofHominoids." Journal of Molecular Evolution. 30:260-266

33. Glowacka H., McFarlin S.C., Catlett K.K., Mudakikwa A., Bromage T.G., Cranfield M.R., Stoinski T.S., Schwartz G.T. (2016). Age-related changes in molar topography and shearing crest length in a wild population of mountain gorillas from Volcanoes National Park, Rwanda. American Journal of Physical Anthropology 160:3-15.

34. Gray, M., Mc Neilage, A., Fawcett, K., Robbins, M., Ssebides, B., Mbula, D., Uwingeli, P. (2010). Censusing the mountain gorilla in the Virunga Volcanoes: complete sweep method versus monitoring. African Journal of Ecology, 48:588-599.

35. Gray, M., Roy, J., Vigilant, L., Fawcett, K., Basabose, A., Cranfield, M., Uwingeli, P., Mburanumwe, I., Kagoda, E., Robbins, M.M. (2012). Genetic census reveals increased but uneven growth of a critically endangered mountain gorilla population. Biological Conservation, 158: 230-238.

36. Grehan, JR. (2006). Mona Lisa smile: the morphological enigma of human and great ape evolution. *The anatomical Record* 289B, 139-157.

37. Gregory, W.K. (1916). Studies on the evolution of the primates, Part II — phylogeny of recent and extinct anthropoids with special reference to the origin of man. *Bulletin of the American Museum of Natural History* 35, 239-355.

38. Gribbon, John and Jeremy Cherfas (1982) The Monkey Puzzle, Pantheon Books, New York

39. Groves, Colin P. (1986), "Systematics of the Great Apes." Comparative Primate Biology? 1:187-217

40. Grueter, C.C., Ndamiyabo, F., Plumptre, A.J., Abavandimwe, D., Mundry, R., Fawcett, K.A., & Robbins, M.M. (2012). Long term temporal and spatial dynamics of food availability for mountain gorillas in Volcanos National Park. American Journal of Primatology. DOI: 10.1002/ajp.22102.

41. Grueter, C.C., Robbins, M.M., Fawcett, K.A., Ndagijimana, F., Stoinski, T. (2013). Anecdotal evidence of tool use in mountain gorillas. Behavioral Processes.

42. Grueter, C.C., Deschner, T., Behringer, V., Fawcett, K., Robbins, M.M., (2014). Correlates of energy balance in wild female mountain gorillas: application of a c-peptide based biomarker. Hormones and Behavior.

43. Grueter, C., Robbins, A., Abavandimwe, D., Vecellio, V., Ndagijimana, F., Ortmann, S., Stoinski, T.S., & Robbins, M.M.. (2015). Causes, mechanisms, and consequences

of content competition among female mountain gorillas in Rwanda. Behavioral Ecology, doi:10.1093/beheco/arv.212.

44. Grueter, C. & Stoinski, T.S. (2016). Homosexual behavior in female mountain gorillas: reflection of dominance, affiliation, reconciliation or arousal? Plos One.

45. Harrison, T. (2012). Apes among the tangled branches of human origins. Science 327, 532-534.

46. Hamburg, D. (1968), 'Evolution of emotional responses: Evidence from recent research on non-human primates', Science and Psychoanalysis, No. 12, pp. 39-54.

47. Habumuremyi, S., Robbins, M.M., Fawcett, K.A., & Deschner, T. (2014). Monitoring ovarian cycle activity via progestagens in urine and feces of female mountain gorillas: A comparison of EIA and LC-MS measurements. American Journal of Primatology, 76, 180-191.

48. Ito, A., Eckardt, W., Stoinski, T.S., Gillespie, T.R., Tokiwa, T. (2015). Prototapirella ciliates (Entodiniomorphida) from wild habituated Virunga mountain gorillas (Gorilla beringei beringei) in Rwanda with the descriptions of two new species European Journal of Protistology

49. Karama J. (2011). Developing an in situ conservation education program in Rwanda. A case study of the Dian Fossey Fund's primary school program around Volcanoes National Park. Journal of the International Zoo Educators Association 47: 46-51.

50. Kralick A., Burgess L., Arbenz-Smith K., Glowacka H., McGrath K.J., Chan K.C., Cranfield M.R., Stoinski T.S., Bromage T.G., Mudakikwa A., McFarlin S.C. 2017. A radiographic study of molar development in wild Virunga mountain gorillas of known chronological age from Rwanda. American Journal of Physical Anthropology. 163(1): 129-147.

51. Laura Boness (2012 March 09)[web post] What do gorillas and humans have in common? Retrieved July 08, 2018 from

http://scienceillustrated.com.au/blog/nature/what-do-gorillas-and-humans-have-in-common/

52. Langergraber, K.E. Prüfer, K., Rowney, C., Boesch, C., Crockford, C. Fawcett, K., Inoue, E., Inoue-Muruyama, M., Mitani, J.C., Muller, M.N., Robbins, M.M., Schubert, G., Stoinski, T., Viola, B., Watts, D., Wittig, R.M., Wrangham, R.W., Zuberbühler, K., Pääbo, S., & Vigilant, L. (2012). Generation times in wild chimpanzees and gorillas suggest earlier divergence times in great ape and human evolution. Proceedings of the National Academy of Sciences, September 25, 2012, vol. 109 no. 39, p. 15716-15721.

53. Madinda, N.F, Ehlers, B., Wertheim, J.O., Akoua-Koffi, C., Bergl, R.A., Boesch, C.A., Akonkwa, D.B.M., Eckardt, W., Fruth, B., Gillespie, T.R., Gray, M., Hohmann, G., Karhemere, S., Dujirakwinja, D., Langergraber, K., Muyeme, J., Nishuli, R., Pulay, M., Petrzelkova, K.J., Robbins, M., Todd, A., Schubert, G., Stoinski, T., Wittig, R.M., Zuberbuhler, K., Peeters, M., Leendertz, F.H., Calvignac-Spencer, S. 2016. Assessinghost-virus co-divergence for close relatives of Merkel cell polyomavirus infecting African great apes. Journal of Virology. 91:11.

54. Marks, J. 2009. Why I Am Not A Scientist. London: University of California Press.

55. Marks, J. 2012. Why Be Against Darwin? Creationism, Racism, and the Roots of Anthropology. Yearbook of Physical Anthropology, 55: 95-104.

56. McFarlin, S., Barks, S.K., Tocheri, MW., Massey, J.S., Eriksen, A.B., Fawcett, K.A., Hof, P.R., Bromage, T.G., Mudakiwkwa, A., Cranfield, M.R., Stoinski, T.S., Sherwood, C.C. (2012). Early brain growth cessation in wild Virunga mountain morillas (Gorilla beringei beringei). American Journal of Primatology.DOI: 10.1002/ajp.22100.

57. McFarland, K. L. 2007. Ecology of Cross River Gorillas (Gorilla gorilla diehli) on Afi Mountain, Cross River State, Nigeria. Ph.D. Thesis, Graduate Center, City University of New York, New Yo

58. Nahayo A., Bigendako M. J., Fawcett, K., Gu Y. (2010) Ethnobotanic study around Volcanoes National Park, Rwanda. New York Science Journal3(5): 37-49.

59. Pilbeam, David. (1996),"Genetic and Morphological Results." Molecular Phvlogenetics and Evolution. 5(1): 155-168

60. Reuters, Carberry, et al (2016 April, 04) "A gorilla named Susie illustrates genome similarities with humans" Retrieved July 08, 2028 from 1. https://www.washingtonpost.com/national/health-science/the-gorilla-genome-shows-were-very-similar-to-them/2016/04/04/0e30f51a-f80e-11e5-a3ce f06b5ba21f33_story.html?noredirect=on&utm_term=.a7a90ab65a49

61. Remis, M.J. 1997. Ranging and grouping patterns of a western lowland gorilla group at Bai Hokou, Central African Republic. American Journal of Primatology 43: 111-133.

62. Robbins, A. M., Stoinski, T. S., Fawcett, K. A. & Robbins, M. M. (2011). Lifetime reproductive success of female mountain gorillas. American Journal of Physical Anthropology, 146: 582-593.

63. Robbins, M.M., Gray, M., Fawcett, K.A., Nutter, F., Uwingeli, P., Mburanumwe, I., Kagoad, E., Basabose, A., Stoinski, T.S., Cranfield, M.R., Byamukama, J., Spelman, L.H., Robbins, A.M. (2011). Extreme conservation leads to the recovery of the mountain gorilla. PLoS ONE 6(6): e19788. doi:10.1371/journal.pone.0019788.

64. Rosenbaum, S., Silk, J. & Stoinski, T.S. (2010). Male-immature relationships in multi-male groups of mountain gorillas. American Journal of Primatology, 71, 1-10.

65. Rosenbaum S, Hirwa JP, Silk JB, Vigilant L, & Stoinski T.S. (2015). Male rank, not paternity, predicts male-infant relationships in mountain gorillas. Animal Behaviour, 104, 13-24

66. Rosenbaum, S., Vecellio, V. & Stoinski, T.S. (2016). Mixed-sex, multi-age coalitionary attacks in wild mountain gorillas. Science Reports

67. Rosenbaum S, Hirwa JP, Silk JB, Vigilant L, & Stoinski T.S. (2016). Infant mortality risk and paternity certainty are associated with postnatal maternal behavior toward adult male mountain gorillas (Gorilla beringei beringei). PloS one 11, no. 2 (2016): e0147441.

68. Rosenbaum S, Hirwa JP, Silk JB, & Stoinski T.S. (2016). Relationships between adult male and maturing mountain gorillas (Gorilla beringei beringei) persist across developmental stages and social upheaval. Ethology, 122, 134-150.

69. Roy, J., Gray, M., Stoinski, T.S., Robbins, M.M., Vigilant. L. (2014). Fine scale genetic analysis suggests further dispersal in male than female mountain gorillas. BMC Ecology 2014, 14:21. http://www.biomedcentral.com/1472-6785/14/21

70. Rushmore, J. L., Caillaud, D., Richard, J., Stumpf, R. M., Meyers, L. A., & Altizer, S. (2013). Network-based vaccination improves prospects for disease control in wild chimpanzee. Journal of Animal Ecology, DOI: 10.1111/1365-2656.12088

71. Ruvolo, Maryellen. (1994), "Molecular Evolutionary Processes and Conflicting Gene Trees: The Hominoid Case." American Journal of Physical Anthropolgy. 94:89-113

72. Ruvolo, Maryellen. (1997) "Molecular Phylogeny of the Hominoids: Inferences from Multiple Independent DNA Sequence Data Sets." Molecular Biology and Evolution. 14(3): 248-265

73. Schwartz, Jeffrey (1987), The Red Ape Hypothesis. Houghton Mifflin, Boston

74. Stoinski, T.S., Perdue, B., Breuer, T. & Hoff, M. (2013). Variability in the developmental life history of the Genus Gorilla. American Journal of Physical Anthropology.

75. Strier, K.B., Atlmann, J., Brockman, D., Bronikowski, A., Cords, M., Fedigan, L., Lapp, H., Liu, X., Morris, W., Pusey, A., Stoinski, T.S., Alberts, S. (2010). The primate life history database: A unique shared ecological data resources. Methods in Ecology and Evolution, 1, 199-211.

76. Sciencemuseum.org.uk "What are our closest animal relatives?" Retrieved July 08, 2018fromhttp://whoami.sciencemuseum.org.uk/whoami/findoutmore/yourgenes/where didwecomefrom/whatareourclosestanimalrelatives

77. Shoshani, Jeheskel et al. (1996), "Primate Phylogeny: Morphological vs. Molecular Results." Molecular Phvlogenetics and Evolution, 5(1): 102- 154

78. University of Washington Health Sciences/UW Medicine. "Improved ape genome assemblies provide new insights into human evolution: Better understanding of genetic influences on primate and human brain differences was also gained through comparative organoid models." ScienceDaily. ScienceDaily, 7 June 2018. <www.sciencedaily.com/releases/2018/06/180607141036.htm>.

79. Vigilant, L., Roy, J., Bradley, B., Stoneking, C.J., Robbins, M.M., & Stoinski, T.S. (2015). Reproductive competition and inbreeding avoidance in a primate with habitual female dispersal. Behavioral Ecology and Sociobiology, 69, 1163-1172.

80. Will, D.(2016 March 31) "A gorilla named Susie illustrates genome similarities with humans, Retrieved July 08, 2018 from https://www.reuters.com/article/us-science-gorillas/a-gorilla-named-susie-illustrates-genome-similarities-with-humans-idUSKCN0WX2UV

81. Wright, E., Grueter, C.C., Seiler, N., Abavandimwe, D., Stoinski, T.S., Ortman, S., Robbins, M.M.. (2015). Energetic responses to variation in food availability in the two mountain gorilla populations (Gorilla beringei beringei). American Journal of Physical Anthropology, DOI: 10.1002/ajpa.22808.

82. Yunis, Jorge and Om Prakash. (1982) "The Origin of Man: A Chromosomal Pictorial Legacy." Science, 215:1525-1529

83. Zev N. Kronenberg, Ian T. Fiddes, David Gordon, Shwetha Murali, Stuart Cantsilieris, Olivia S. Meyerson, Jason G. Underwood, Bradley J. Nelson, Mark J. P. Chaisson, Max L. Dougherty, Katherine M. Munson, Alex R. Hastie, Mark Diekhans, Fereydoun Hormozdiari, Nicola Lorusso, Kendra Hoekzema, Ruolan Qiu, Karen Clark, Archana Raja, AnneMarie E. Welch, Melanie Sorensen, Carl Baker, Robert S. Fulton, Joel Armstrong, Tina A. Graves-Lindsay, Ahmet M. Denli, Emma R. Hoppe, PingHsun Hsieh, Christopher M. Hill, Andy Wing Chun Pang, Joyce Lee, Ernest T. Lam, Susan K. Dutcher, Fred H. Gage, Wesley C. Warren, Jay Shendure, David Haussler, Valerie A. Schneider, Han Cao, Mario Ventura, Richard K. Wilson, Benedict Paten, Alex Pollen, Evan E. Eichler. **High-resolution comparative analysis of great ape genomes**. *Science*, 2018; 360 (6393): eaar6343 DOI: 10.1126/science.aar6343

Author: Nsesheye Nkota

Note From the Author:

Reviews are gold to authors! If you've enjoyed this book, would you consider rating it and reviewing it on www.Amazon.com?

How do I write a book review on Amazon?

To submit a review:

1. Go to the product detail page for the item on Amazon.com

2. Click Write a customer review in the Customer Reviews section.

3. Rate the item and write your review.

4. Click Submit.

If you have any question don't hesitate to contact me through: nzeshely@gmail.com

Tel: +255-685-045-910

Arusha, Tanzania

www.ingramcontent.com/pod-product-compliance
Lightning Source LLC
Chambersburg PA
CBHW070522220526
45467CB00002B/803